# Something big is afoot in Colorado

Tuesday, March 01, 2016

**Frank first encountered Bigfoot while hunting in the Colorado mountains when these large footprints in the snow (above right) caught his attention and led to his subsequent hunt for further evidence of Bigfoot. Frank made his own tracks in the snow (above left) to indicate the size. Bigfoot sightings have taken place in Colorado for more than 200 years. This (actual) state highway sign was erected on the Pikes Peak Highway more than 20 years ago after a local resident produced evidence of the creature.**

*Article and photo by Amy Shanahan; additional photos courtesy of Jason Frank*

Jason Frank is a high energy, entertaining gentleman who has a successful local business called Wildman Wellness, where he practices Muscle Activation Techniques (MAT) which help improve and correct muscular weakness. Frank is a former registered nurse and served as an infantryman/combat lifesaver in the United States Army 82nd Airborne Division. Oh, and in his spare time he follows groups of wild animal creatures that most of us have come to refer to as Bigfoot.

"I know people initially think I must be crazy, but I'm stubborn and enjoy proving them otherwise," explained Frank.

Frank's hunt and subsequent encounters with Bigfoot creatures began in 2007 when he was hunting turkeys in the Colorado mountains. Frank has been an avid hunter, fisherman, and outdoorsman his entire life and was hunting in a remote forested area. Quite unexpectedly, Frank came across unusually large footprints in the snow (pictured above) and he was astounded by the size of the prints and the size of the tread between the prints. Frank photographed the prints and sent them to a renowned Bigfoot expert at Idaho State University, Dr. Jeffrey Meldrum. Meldrum felt the photo was indeed evidence of Bigfoot, and thus began Frank's quest to acquire further evidence of what he believes to be a very real species of creature.

Over the past eight years, Frank has had numerous encounters with Bigfoot creatures in an area he discovered near Steamboat Springs. According to Frank, the creatures are nocturnal so many of his encounters have taken place during the night. He has had his tent surrounded by the creatures, his camping coffee pot has teeth marks in the lid, he has a plastic bag which was poked by a creature, and he has numerous photos of footprints, handprints and other evidence of their existence.

Frank's most up close and personal encounter with a Bigfoot creature occurred three years ago on an expedition with a large group in New Mexico, when his group saw a creature walk and run through the forest. The group was made up of members of the Bigfoot Field Researchers Organization (BRFO) which is a scientific research organization dedicated to obtaining conclusive evidence of Bigfoot. Frank and many other Colorado residents are members of this organization.

Frank's most eerie evidence of Bigfoot is the multitude of sound recordings that he has made. As much of their movement takes place at night, Frank records while sleeping and he has been often disturbed by guttural howls, cries and grunts.

Frank believes that the creatures do not mean any harm, but they are curious. His experience has been that they observe humans and explore and taunt the humans only when they feel they are not threatened. "You have to be careful though," explained Frank. "You are dealing with a wild animal so you have to be respectful and the fear is always there."

Frank has had the opportunity to make believers out of many of his closest friends and family. Frank's wife was skeptical until she heard and witnessed the creatures first-hand. Numerous friends have accompanied Frank on his camping trips and have had encounters of their own. "Speaking out about Bigfoot subjects people to a lot of ridicule, but I'm very open to talking about it with anyone who is interested," stated Frank. "I've made some wonderful friends out of people who were skeptical and are now believers. I even named my business after the creatures so that people are aware upfront that this is a passion for me."

# "I know people initially think I must be crazy, but I'm stubborn and enjoy proving them otherwise." – Jason Frank

Bigfoot sightings have been around for hundreds of years, as even miners in the 1800s in Colorado claimed to have seen unusually tall and hairy creatures. Numerous claims have been made throughout the world over the years, although scientists and the general public remain skeptical and hoaxes have been discovered. In recent years, scientists have begun studying DNA samples and looking more closely at evidence collected by people like Frank and other members of BFRO. Animal Planet even dedicated an entire weekly show "Finding Bigfoot" to the subject. Frank believes that there is a great deal of fear involved, which leads to skepticism. His view is quite the opposite. "In my opinion, this is the most important scientific discovery of all time. These creatures teach us about our roots."

Frank is extremely passionate about his connection with these creatures. "Having these experiences really makes you look at the world differently. It makes the world full of wonder and you begin to see things as a child would. My encounters with these creatures have changed my life for the better."

To learn more about Bigfoot, Frank suggests the following resources: Sasquatch Outpost which is based in nearby Bailey, CO,

**Colorado Bigfoot – hundreds of sasquatch sightings draw attention to elusive creature hiding in Colorado's remote mountains**

Sasquatch, Bigfoot and Yeti. It goes by many names and has been said to call Pike National Forest home. Passersby, visitors, park rangers and investigators alike have told their tall tales of seeing the elusive, hairy creature wandering the woods along the twists and turns of the Pikes Peak Highway. Sightings first reported in the area date back to the 1800s.

After multiple reports of encounters, a Bigfoot crossing sign was finally placed along the Pikes Peak Highway around 1990. In 2013, an episode of Animal Planet's *Finding Bigfoot* was filmed in the region and local researchers with Sasquatch Investigations of the Rockies have snapped possible Bigfoot imagery. A local artist even carved the creature's likeness in wood with an interpretive sign that visitors can view at Crystal

Reservoir, a popular stop along the Pikes Peak Highway.

Pikes Peak National Forrest/Muller State Park Divide Colorado. John Kuykendall

Paranormal Book Author and Cryptid Investigator, along with his Camera man Xavier Quinoens, has documented evidence in the Pikes Peak National Forrest of Bigfoot nests and 18 by 5 1/2 inch wide foot prints. Just 1 mile NW of this sign.

Possible Bigfoot nest found in the Pikes Peak Nation Forrest on Bigfoot Highway

Camping and picnic areas at Mueller State Park in Divide we found what could be nests that where built like something had been living in them. As we walked lower down the mountain we also found several other nest where 1000 pound trees were moved up on rocky mountain sides that no human could possibly move. These 1000 pound trees had smaller logs leaned up on these trees and you can see bedding placed inside and empressions of something large that was laying in them. Could this of been a bigfoot? I do not know what else could of made these nest?

There is a pond at the base of these nest and as we know everything must drink. Just to the other side of the pond there is a trail that we visited back in January where we found 18 by 7 in foot prints that were bipedel and they were 6 inches deep, I stood next to the prints and put all my weight down and I way 230 pounds and I could press the snow down about 2 inches. What ever was walking there had to be at least 800 pounds. As we returned to this area 3 months later we continued down the path of the foot prints we found in January and about 200 yards down the trail we came across another print that looked liked something crossed the 6 foot wide trail from one side to the other, because

we got a print that stepped on the embankment that led to the wooded area. This print was 12 inches long and 6 inched wide. You can see on the next page. My first instinct was to say it was a bear but there are no bears in this area. We then followed the direction of the track about 30 yards into the wooded area Arron decided to do some tree knocks that bigfoots are know to use to communicate so he did three knocks and a few seconds later we heard three knocks back, so he did it again and we got three knocks back. Then a few seconds later you can hear something heavy walking like it was right next to you and something is going to come up on you. It was loud and our camera man that was 30 yards back on the trail was hearing it also. We looked at each other like "what the hell is that? Our hearts started beating fast like we knew we were going to encounter this 9 foot creature at any moment. But nothing? We could hear it but we could not see it anywhere, but it kept walking then it stopped. Then a growl came then it went into a howl and what I have heard over the years of the sounds that bigfoots make, this was what it sounded like to me. We decided to go back the way we came before we encountered something we may not of wanted to.

On our walk back up the trail to get to the car we could hear something walking in the brush behind us, like something was following us. as we continued up the trail Arron heard it again but this time it was in front of us walking. We shined our lights but again we could not see anything?

I spoke to a couple of park rangers about our experience and showed them some of the footage we took and they were stunned to see what we captured on film. They have never seen these nest before since no one goes that deep into the Forrest.

It got them thinking that there could be something in these Forrest. I told them that I

would be back to do a over night investigation and camp right in the middle of this activity and see if I can get in the way of what ever is making these nest, and that might cause them to try and scare me out of there territory. Then I can see for myself if these creatures exist?

On Friday, April 1, 2016, sightings were taken to a whole new level when Pikes Peak Highway drivers claimed to have seen the creature use the Bigfoot crossing in front of them. Videos and photos can be found and shared by using the #BFFCOS hashtag.

Whether real, imagined or fabricated, the stories continue to be told and evidence continues to be shared about this over sized, but friendly Colorado Springs resident.

It's a question that never grows old: Is Bigfoot out there? Commercials, movies, television shows, documentaries, books, and most important of all, photos, have focused on a legendary primate-type critter who stands between seven and ten feet tall and enjoys lumbering around in the woods. He is generally described as smelly but gentle-appearing, walking with slow-moving gracefulness or loping through a field. In some instances he looks casually at the observer but doesn't appear the least bit interested in the humans who spot him.

At night, some claim, you can hear his haunting screams echoing through the remote backwoods. Some have said he grunted at them during an encounter. Others say he is responsible for making knocking noises with large pieces of wood in the forest. Still others have given reports of having rocks thrown at them and witnessing feasts by the beast on wild game. Depending on the region, this giant ape-like being goes by Sasquatch, Yeti or, in Colorado, Bigfoot.

Bigfoot sightings have in fact been going on for centuries all over the world, including the United States. According to the official Bigfoot Field Researchers Association, Hawaii is actually the only state without one. Delaware and Rhode Island rank the next lowest with only five sightings each. Washington State ranks first with 577 sightings since 1996. California comes second with 427, and Oregon has had 235.

Indeed, the Pacific Northwest was where the first Bigfoot sighting in America occurred, clear back in 1811. But in the great ranking of the most sightings, it is interesting to note that Colorado takes 8th place (the state is actually tied with Georgia at 115 sightings). The first time anyone reported seeing a Bigfoot in Colorado was in Jackson County, when a hunter watched two of them stalk an elk way back in 1926. For many years Teller and Park counties, nestled next to each other and surrounded by plenty of remote forests, took first honors above the rest of the state with numerous sightings. Increased populations and mining activity in Teller County especially have changed the numbers only slightly in recent years; these days Park County leads with nine sightings, while Teller, Lake and Conejos counties have each had eight. Nearby El Paso County follows with seven sightings.

"Bosh!" say some who staunchly deny the big hairy guy exists at all. Enough reports have surfaced, however, to merit looking into the matter further. Take 1972, for example, when a couple hiking in the Lost Creek Wilderness observed a creature squatting near a pond. "When he stood up erect and looked at us we knew it was not a bear," stated one of the witnesses. Although the couple was unable to obtain a photograph on subsequent visits, "Unsolved Mysteries" television show covered the incident in a segment during the early 1990's.

Closer to Cripple Creek were two sightings in the 1970's, one during the day by two brothers hiking and another by a young girl camping with her family. Both happened on the west side of Pikes Peak. In the latter instance, the creature was peering into a camper trailer when the girl awoke and came face to face with it through the screened window. "I quickly pushed the curtain closed and laid there completely paralyzed with fear," she recalled in later years. The woman also remembered hearing a "chatting" noise that first awoke her, and that the next morning the apples and potatoes had been stolen from the back of the family truck. During that same time, other hikers noted large footprints in the snow on the backside of the Peak.

Just a few years later, in 1981, a mine watchman near Cripple Creek was badly shaken when he spotted a tall, two-legged creature near one of the mine buildings. In 1984, another young girl spotted the long-armed hairy man walking amongst some cows along Phantom Canyon Road. More tracks were found near the Cripple Creek watershed in 1986. And in 1987, two Florence boys working at a donut shop watched "a very large hairy looking thing" sauntering down the main drag on two legs in the dead of night.

By then, Roger Patterson and Robert Gimlin's famous film documenting Bigfoot for the first time had been debated over for some twenty years. Even today the seconds-long footage of a hairy beast strolling along a creek in 1967 infatuates researchers everywhere. Gimlin recently theorized the incident might have been a hoax, especially since hundreds of such have been perpetrated over the last 40 years.

In 1987, for instance, Green Mountain Falls resident Dan Masias claimed to have seen two Bigfoots booking down the road in front of his home. When unidentifiable hair was found on the door of a home that was broken into, the story began making international headlines. Masias' last sighting was in 1992, but rumors circulating that he confessed to faking the whole thing slowed down the amount of witnesses coming forward for a few years.

Once things cooled down, the Bigfoot sightings made a reprise. Cadets at the United States Air Force Academy reported seeing and hearing a Bigfoot from Jack's Valley above the base throughout the 1990's. Four accounts in Teller, Douglas and El Paso counties in 1997—including one along Ute Pass—plus four sightings in as many surrounding counties in 1998, and the big guy was back on top. In December of 1999 a trucker spotted one south of Colorado Springs. During the fall of 2000, people reported hearing "eerie human-like calls" and finding footprints in the Pikes Peak National Forest. Also, a tribe of Romanian gypsies near Fairplay beat a hasty retreat after a seven-foot tall creature appeared at the edge of their campsite.

The sightings continued. In May 2001 a woman and two small children living near Lake George distinctly heard an animal emit a blood-curdling "rooster-dog" howl outside their home. A month later, when a nine-year-old boy spotted a Bigfoot at the first rest stop along the road to the top of Pikes Peak, officials had already half-jokingly posted a sign warning of Bigfoot sightings. Finally, a hunting guide and three other people heard the rumble of an unidentifiable animal, heard something knocking logs together and found large but melting tracks in the snow off Gold Camp Road near Victor in November.

During 2002, reports in El Paso and Park counties told of hearing eerie screams, including one that was in response to a hunter's elk bugle. There were two more sightings in 2005. The first was in January near the Crags Campground, where some mighty large footprints were photographed in the snow. Then in October of 2005, employees of the Arrowhead Gold Course off Range View Road in Douglas County saw a "huge whitish gray figure" peaking at them on Hole #13. And in May of 2006 a hunter saw a Bigfoot walking in the hills west of Fort Carson.

More recently, in August 2010 a woman delivering newspapers west of Buena Vista during in the early morning hours was startled to see "a large upright dark figure" cross a two-lane highway in just three steps. And in May of 2012, two women on an evening hike near Bailey watched as a creature measuring seven to eight feet tall and standing upright ran from them into the woods. And for every sighting reported on the BFRO website, numerous others – such as huge tracks found in the snow near Cripple Creek during the winter of 2012 – do not get reported.

Going by all accounts, the Bigfoots of the Pikes Peak region have been seen, heard, or left their tracks at all times of the year. They seem to favor steep and forested terrain. Upon seeing humans they usually exhibit a gentle curiosity before moving on, rarely venturing nearer. Now and then they let out a screaming howl, sometimes in answer to the call of another animal. And, they tend to be a bit smelly. Most of the many websites covering the phenomena advise to be at the ready with a camera if you see one. None of the sites say what to do if you really do see one, but those who believe ought to have some extra fun hiking and camping this summer.

LOCATION DETAILS: first visitor parking lot on the Pike's Peak Highway, right turn (going uphill)

NEAREST ROAD: Pike's Peak Highway

OBSERVED: My family was on vacation in Colorado Springs, Co, last summer. We were driving up Pike's Peak on the main road, and turned off onto the first rest area parking lot.
The region was composed of large and small rock formations and was forested, primarily with pine trees. A stream was across the street. It is an area that is not inhabited by human, except for the traffic on the road: tourists visiting Pike's Peak.
I was standing on top of a large rock formation, looking down on my dad, who was about to take my picture with his camera.
I shouted, "Dad, look!" as I saw what I immediately believed to be a bigfoot 100 feet behind him. My dad did not turn around and look at the bigfoot, because he wanted to take my picture.
The bigfoot was in motion, black, and was quickly out of sight.
I know what bears look like, because I did a report for school on bears last year. This was not a bear.

ALSO NOTICED: No sounds, no smells.

OTHER WITNESSES: I, Nevan, 9 years old, was the only witness.
Just before I saw the bigfoot, my mother, brother, and two boy cousins began to walk around in the forest, and climb on the rock formations. Perhaps our activity caused the bigfoot to want to leave the area, and he or she was attempting to run away from us.

OTHER STORIES: My family and cousins got back in our car to continue our trip up Pike's Peak.
To our surprise, after we went a very short distance, not even a quarter mile, there was a green, official sign, and we read in amazement that it was displaying a Bigfoot picture. All the way up Pike's Peak highway, there were placed "wildlife" signs about what tourists may see in that particular zone.
The sign, placed by official people was of Bigfoot.
I repeat, we did not come upon this sign until after I had seen Bigfoot with my own eyes.

TIME AND CONDITIONS: 11:15 a.m.
partly cloudy (clouds are usually covering the top of Pike's Peak)

ENVIRONMENT: pine forest, numerous rock formations, hilly, remote (from humans, foothills of Pike's Peak, freshwater across the street

YEAR: 2001

SEASON: Summer

MONTH: June

DATE: 30

STATE: Colorado

COUNTY: El Paso County

Follow-up investigation report:

> Twice I spoke with the witness, Nevan, and his brother Nick, now 12 and 14 respectively. The parents were unavailable for interview. I spoke with the boys separately and neither had any difficulty recalling details of the event.
>
> They came from Minnesota to vacation with their cousins who live in Colorado Springs. The sighting occurred at the Crow Gulch Picnic Ground on FS 334, Pike's Peak Toll Road. Nevan was up on a rock outcrop looking down on his father who was taking a picture of him from below. This is the view from the spot where Nevan stood:
>
> I queried Nevan about the size and shape of the animal, especially trying to elicit from him what is *different* from a bear, that is, why he's confident it wasn't a bear that he saw. This is an excerpt from our conversation:
>
> NH: "It looked pretty big."
>
> KS: "Big like a horse? Big like Hulk Hogan? Or big like your dad? Was it big and thick or big and skinny?"
>
> NH: "It was big and thick, like Hulk Hogan. It was bigger than a black bear."
>
> KS: "Did you see arms? Legs? Did it have shoulders or knees?"
>
> NH: "I saw it from kind of above it. I didn't see its legs, but the shoulders weren't on its back like a bear's are. The shoulders didn't move on its back as it ran like a bear's do."
>
> KS: "What did it move like? Was it bouncy or smooth? Fast or slow?"

NH: "It moved smooth and fast. It didn't bounce up and down like a bear running. I saw it for 3 seconds, looked at my dad telling him to look, and when I looked back it was gone."

KS: "Where did it come from? Where did it go? Why do you think it was running there?"

NH: "I never thought about why it was there. That's a good question. I don't know where it came from and then it was just gone."

The grass the animal moved through is taller than it appears in the photograph. Most is around 16" tall with much of it reaching 2 1/2'. Coupled with the witness' elevated position, I'm confident this explains why he didn't see the animal's legs or feet. I inspected the site with this specific question in mind.

In addition to trees and grass, as described in the report, there is also more dense moist-site vegetation following the draw north, upslope, toward the location of Report 1359 1- 2 miles distant. In the opposite direction, the draw meets Cascade Creek and its riparian character, which continues upstream toward the location of Report 819 at its headwaters 5-6 miles distant.

Beyond topology, there is no indication that the same animal was involved in the three encounters. It's interesting, though, that the Bigfoot Crossing sign on the road near Nevan's sighting is actually where an animal would cross to move between the locations of reports 1359 and 819 if it were traveling along the creek.

# Teller County Mine Bigfoot sighting (Teller County, Colorado, 1970's)

A sighting in *Teller County, Colorado* (near *Colorado Springs*, the location of several notable Bigfoot sightings) left a grown man in tears. A nephew of the man recalled seeing his uncle, who had just completed the nightly rounds in the Cripple Creek mine ponds to make sure all equipment was functioning properly, scramble into a secluded miner's building, burst into tears, then calling his father from the cabin phone to report seeing "a creature larger than anything he'd ever seen before".

> "I was 13 at that time, and I had never seen or heard my uncle cry. I am 29 now and I have never seen or heard him cry since that time."

The sighting so upset the man, to this day he refuses to discuss the encounter with researchers.

# Bailey Bigfoot sighting (Bailey, Colorado, 2012)

The *Bailey Bigfoot sighting* occurred at approximately 7:30 PM on the evening of May 29, 2012. Kate Murphy and an unnamed neighbor, were walking a path in the wilderness area near Bailey, Colorado, about a mile off of *US Highway 285*, when they heard a loud tree branch snap off in the distance. As they turned their heads towards the noise, they saw an extremely large figure running through the woods a few hundred feet away. It was running incredibly fast, on two long hairy legs, with a very long stride. Neither of them could clearly see the top half of the body, but it appeared to have been leaning forward while running. Familiar with the area, they concluded it was not a deer, elk, bear or moose. The women noted that the creature's stride was very long and that it had no problems navigating through the trees. The creatures hair was dark with signs of graying throughout.

By 2012, Bigfoot research had become popular and an experienced research team immediately descended on the area towing the two eyewitnesses in hand. Using precise positioning of the participants (both the two women and the unidentified creature) and advanced metering equipment, the researchers were able to determine that the creature, when bent over, was about 7' 2" tall and situated about 75 yards from the eyewitnesses. They estimated that if the creature had stood upright, it would have been well over 9 feet tall in height.

    The researchers combed the area and found several ground anomalies including odd depressions where the creature stood amongst the trees. Deep footprints were found embedded in the ground, measuring 22 inches long by 8 inches wide and spaced approximately 10 feet apart. They also found, and photographed, broken tree branches and felled trees along the path the creature took.

The convincing testimony of the two women along with the ample physical evidence obtained by researchers has made the Pike National Forest sighting one of the best bigfoot sightings to date.

    We are continuing are search for these elusive creature's and we will investigate every possible sighting in the Pikes Peak National Forrest. We have heard of bigfoot abductions and bigfoot rapes of humans? These are new reports that the news channels are reporting so we might want to look into this. These creatures are changing the way they encounter us. Other reports that UFO's are seen in area that bigfoot sightings take place. Here is a story that took place along the platte river 33 years ago that we will call Jeff called into Coast to Coast am to report.

    "I am an engineer and I'm an eyewitness to a Bigfoot sighting here in Colorado, along the Platte River. This was about 33 years ago, so it was 1984 on Thanksgiving and a friend of mine was with me...so he also saw it. We had an extended look at this what appeared to be to us like a 7-foot guy in a gorilla suit that was intentionally stamping the ground and breaking limbs as if to scare away predators or something. We had a perfect view. We were sitting on a ridge looking over a pond at a row of trees that was along a ditch on Slaughterhouse Gulch. We saw a Sasquatch or something, you know, we were trying to convince each other that it was a guy in a gorilla suit but that didn't work and we were scared to death. So we sat there petrified as this thing broke through the trees and stomped on the ground and walked along this bank. It was a good 200 meters or more that we were in eyesight contact, right. And we both we just sat there and stared.

    There had been a whole wave of Bigfoot sightings that year. In fact, a girl got kidnapped on Cherry Creek which is one of the tributaries of the Platte River and she swore that it was Bigfoot that had abducted her and carried her over his shoulder. There was another older girl, a teenage girl, who told a similar story. And when we went to investigate the next day we found footprints but they were in the snow. The same thing at the Cherry Creek sighting. We went there the next day and we found footprints but they were in the sand. It had snowed and the snow was melting and the footprints were disappearing as we saw them but they appeared to be about 14 to 16 inches.

There was another sighting that was accompanied by UFO sightings which was at another tributary of the Platte River which is now Chatfield Reservoir and there were deaths involved in that one. Then there was another sighting within a few days of that at a trailer park in the Wolhurst, the previous Wolhurst Country Club which is also along the Platte River. This girl claimed that she was out for a walk, something was rustling through the bushes and whatever it was grabbed her and threw her over its shoulder and she showed us the jacket she was wearing and it was covered in this really stinky weird hair. It was not like fur. It was more like hair but it wasn't really like hair. They were thick, thicker than a human hair. This was almost 33 years ago now. I plucked a couple of the hairs off of the jacket which smelled really bad. It smelled like wet bull or whatever. We were pretty interested at the time because there was a missing girl and we went to go on a search party and they found her just as we got there. The search and rescue team had found her and she was fine."

There may be something that we are not looking for or we may be missing? If these creatures exist and are so elusive maybe there is something else something that does not want us to catch them. Very little DNA has ever been found, no bones or carcass have ever been found? But reports are coming in all the time by people who don't believe in them. They are seeing something? We did not see what they did so we can't say that the people reporting these sightings are liars. The Patterson and Gimlin film to this day has never been proven a hoax? People can say that it was a man in a suit, we could all say that, but there is more evidence showing that bigfoots exist then someone just saying that they don't. People who say they don't have not done the research, it is just easier for them to say they don't. Me myself I am not out to prove anything to the skeptics, I am out to prove it to myself. It doe's not matter to me if people do not believe me that is there problem not mine. Ever since the 1974 film came out The Legend of Boggy Creek, I have always wondered if this creature called Sasquatch existed? I bought everything I could find on the legend and I would read and watch it over and over again to get a better understanding of this creature. I said all of these people can't be lying about what they saw. One day I would search for myself and maybe I would get lucky and see one for myself? Now as a paranormal researcher I have turned to the cryptid side of the paranormal. If bigfoot is so elusive then to me that classifies it as paranormal. Something that has not been governmentally proven.

Then there is this giganticus pithecus The Largest Ape That Ever Lived. Sometimes, in evolution, the bigger they are, the harder they fall.

And *Gigantopithecus* was pretty darn big. Fossils indicate it stood as high as 10 feet (3 meters) and weighed up to 1,100 pounds (500 kilograms)

If you're an animal, there are advantages to being gigantic. You're less vulnerable to predators, and you're able to cover a lot of territory when looking for food. *Gigantopithecus* thrived in the tropical forests of what is now southern China for six to nine million years.

But around 100,000 years ago, at the beginning of the last of the Pleistocene ice ages, it went extinct—because in the changed climate its size had become a fatal handicap, a new study suggests.

"Due to its size, *Gigantopithecus* presumably depended on a large amount of food," explained Herve Bocherens, a researcher at Tübingen University in Germany, in a press statement. "When, during the Pleistocene, more and more forested areas turned into savanna landscapes, there was simply an insufficient food supply for the giant ape."

*Gigantopithecus*, a fruit-eater, failed to adapt to the grass, roots, and leaves that became the dominant food sources in its new environment. Had it been less gigantic, it might have endured somehow. "Relatives of the giant ape, such as the orangutan, have been able to survive despite their specialization on a certain habitat," said Bocherens, because they have "a slow metabolism and are able to survive on limited food."

A Rule … With a Giant Exception

The rise and fall of *Gigantopithecus* illustrates why, over time, size can yield diminishing returns. "There are short-term advantages that come with being bigger, but it also brings long-term risk," says Aaron Clauset, a computer scientist at the University of Boulder, who has studied the body sizes of thousands of species spanning two million years of the fossil record.

It's not just that a bigger body requires more food, says Clauset. It's that "as you get bigger, you tend to have fewer children. That means your population tends to be smaller and more sensitive to fluctuations."

As a result, changes in weather or climate that threaten a food source can reduce a large-bodied species' numbers to the point of "demographic death."

In fact, Clauset found that extinction rates increase as a species gets larger in size. That's why goliaths such as *Gigantopithecus* and the giant sloth no longer roam the Earth. Every animal species has an effective upper limit on how big it can become; how close it can get to the edge of the precipice before toppling off into oblivion.

At least that's the case for mammals—dinosaurs were a different story, Clauset acknowledges. Until an asteroid plunged them into armageddon, they were both enormous and successful for tens of millions of years. Why couldn't *Gigantopithecus* do that? "It may be that mammals have higher metabolic needs, converting more of their energy intake into heat, because they're warm-blooded," Clauset says

    Now could this giant ape that we know existed be considered a bigfoot? If this type of ape could of lived and people saw it and even studied it close up, then could this be were the root of bigfoots came in later? It would have been the same size and weight and it looks like a bigfoot. Could somewhere down the Geno factor had bigfoot evolved from this ape? It could be possible. When people saw this ape 100,000 years ago they were most likely calling it some kind of bigfoot in there day. Look how we changed from the day of the caveman. Our geno factor changed and our looks changed. From ape to man.

*Gigantopithecus*

Evolussion has taken it turns evolving us to what we are today. Could the same thing also happened to bigfoot from this *Gigantopithecus to this 9ft 900 pound giant that we are seeing today evolved just like we did? That looks just like the Gigantopithecus?*

*Scientific Theory . A few years back there was a DNA sample tested.* Bigfoot is people! At least that's according to a new five-year study of the creatures purported DNA by a prominent Bigfootologist. "Genetically, the Sasquatch are a human hybrid with unambiguously modern human maternal ancestry," reads a statement released last weekend by former veterinarian Melba T. Ketchum, the lead researcher of the study. "Researchers' extensive DNA sequencing suggests that the legendary Sasquatch is a human relative that arose approximately 15,000 years ago." Yes, that would mean that Bigfoot is more man-ape than ape-man. However as the hominids are notoriously reclusive, if not entirely fictional — there has never been a single confirmed sighting — it's unclear whether we will need to extend an invitation to our Sasquatch relatives for Easter brunch. For her study, Ketchum "sequenced 20 whole mitochondrial genomes and utilized next generation sequencing to obtain three whole nuclear genomes from purported Sasquatch samples." As her team interpreted their findings, the Sasquatch is a human hybrid with mitochondrial DNA identical to human mitochondrial DNA and nuclear DNA that is of "novel," or non-human, sequence. To hark back to high school biology for a moment: mitochondrial DNA is inherited from the mother, while nuclear DNA mixes genetic material from both parents. That means that according to Ketchum's study, Sasquatch's parents were a human female and some unknown third species, a "novel non-human" male.

At this point it's probably important to note that the study has not yet been peer reviewed and Ketchum has thus far refused to release her data, explain her methodology or say where she got the "Sasquatch DNA samples" in the first place. Also, according to Houston *Chronicle* science writer Eric Borger, Ketchum has credibility issues of her own: her company, DNA Diagnostics, has received more than two dozen customer complaints and gets an F from the Better Business Bureau. Oh, and those mysterious third-species males who were supposedly picking up human women on some kind of proto-Craigslist? According to a blogger and Bigfoot enthusiast named Robert Lindsay, earlier drafts of Ketchum's study claimed they were angels The scientific community remains, unsurprisingly, dubious."The bottom line is this,"Yale neurologist Steven Novella wrote at NeuroLogica Blog:"Human DNA plus some anomalies or unknowns does not equal an impossible human-ape hybrid. It equals human DNA plus some anomalies."

For her part Ketchum, a Texas veterinarian who claims among her bona fides"27 years of research in genetics, including forensics." wants Sasquatch to be immediately afforded civil liberties and protected by state and federal governments as an indigenous people. "Genetically, the Sasquatch are a human hybrid with unambiguously modern human maternal ancestry. Government at all levels must recognize them as an indigenous people and immediately protect their human and Constitutional rights against those who would see in their physical and cultural differences a 'license' to hunt, trap, or kill them," she writes. We guess that would exclude a blimp hunt.

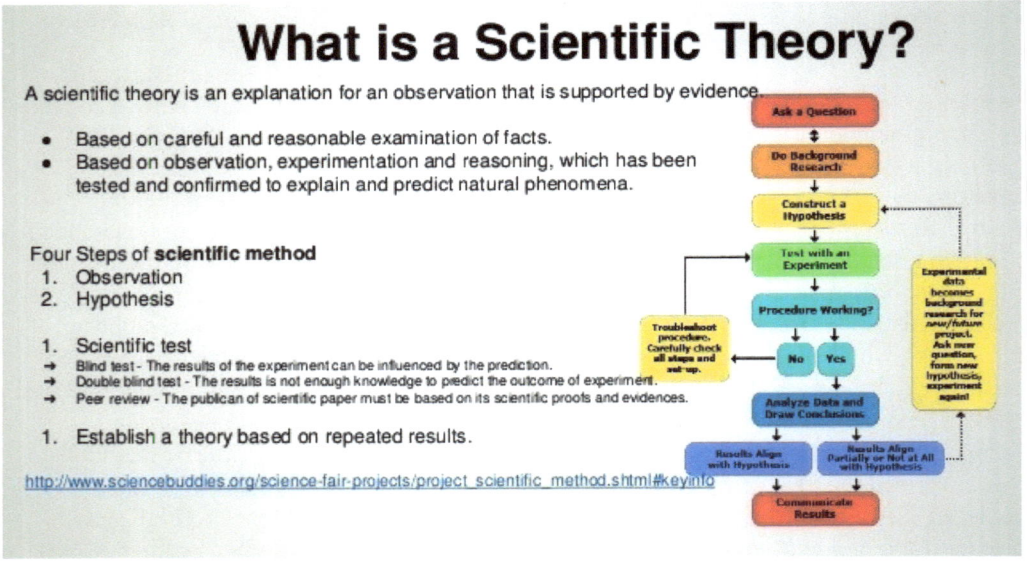

So according to the DNA evidence that was taken from a known bigfoot sample then Scientific Theory would come into play to say that bigfoot exist. So why is science still not excepting it today?

Does **Bigfoot** exist? What about the **Loch Ness monster**? Or the **Yeti**? Or **Mokele Mbembe**, the Congo dinosaur?

There's ample circumstantial evidence for all these creatures: eyewitness accounts, blurry photographs, mysterious footprints. For many cryptozoologists—the people who search for legendary animals—that evidence is enough to confirm a monster's existence.

But it will take more than shadowy sightings to convince [Daniel Loxton](#) and [Donald R. Prothero](#) that Bigfoot or any of the other monsters are real. What Loxton and Prothero want is scientific evidence. In their new book, [*Abominable Science! Origins of the Yeti, Nessie, and Other Famous Cryptids*](#), they analyze the history of mythic beasts and the clues to their existence.

Loxton and Prothero come at cryptozoology from different directions. Loxton, a staff writer for [*Skeptic* magazine](#), was an ardent believer in monsters as a kid, having spotted a Bigfoot print in the woods and a pterodactyl winging over his backyard. (Now, he suspects the print was a prank and the pterodactyl was a [great blue heron](#).) Prothero is a paleontologist, who is also trained in biology and geology. He has written over 250 scientific papers and 28 books, including five textbooks on geology.

National Geographic's Rachel Hartigan Shea spoke with the two authors about bringing skepticism and science to the study of cryptids.

### First of all, what is a cryptid?

DP: A cryptid is any animal that has never been described by science, usually something very unusual along the lines of a Loch Ness monster or Bigfoot, something that stretches the limits of what is scientifically plausible.

DL: It's based on the word cryptozoology, which means hidden life or animals. It implies a creature that's been recorded through folklore, something that we have reason to suspect exists.

### What can science tell us about cryptids?

DP: The first thing, of course, is that a cryptid can't be a single animal. If there's one of them, there's got to be many of them. You can talk about their population density, the size of range they should have based on their estimated body size. All of that tends to weigh against them being real because they should have had huge ranges, and they should have been spotted a long time ago if they really did exist. And then there's other aspects, like geology, something you never hear the cryptozoologists mention. All the lake monsters, not just Loch Ness but the ones here in North America, in [Lake Champlain](#) and [Lake George](#), were all under a mile of ice 20,000 years ago. The cryptozoologists never asked the question, "Well, how did the monster get in the lake if the lake was completely under ice, the lakes are all landlocked, and there's no way for a marine creature to get there at all?" Those are all things that are not news to geologists, they're not news to biologists, but they're apparently news to cryptozoologists.

### All the cryptids that you discuss in the book – Bigfoot, the Yeti, the Loch Ness Monster, Mokele Mbembe – are very similar to things that exist or existed in the past: bears, primates, plesiosaurs, sauropods. Why the similarity?

DL: In some cases I think it's because they are the same. Bears are often associated with ogres or wildmen in folklore because they're pretty humanlike. Once that folklore is underway, you have the opportunity for people to make these misidentification errors where they see a bear and think it might be a bigfoot. (Read a National Geographic magazine story about [Europe's wildmen](#).)

DP: These animals look like something familiar to us because the myths grow around whatever we've already just seen. Daniel pointed out in the book that the Mokele Mbembe myth emerged right about the time that large sauropod skeletons were first mounted in New York City and illustrated by people like [Charles R. Knight](#). Then lo and behold, someone starts reporting one in the Congo, where it doesn't have any history prior to that.

**So Mokele Mbembe definitely does not exist?**

DP: We have an excellent fossil record of Africa. We have very great confidence that there have been no dinosaurs around in the last 65 million years because we have bones of large animals from Africa of all kinds but they're all mammals. Same goes for plesiosaurs. Worldwide, there are no bones of plesiosaurs in any marine deposit after about 70 million years ago. There are plenty of places where they should show up if they actually lived, but they don't. That to me is not just absence of evidence, that's very strong evidence that they don't exist.

**That sentence -- the absence of evidence is not the evidence of absence – occurs a lot in the book.**

DL: It's a really good thing for people to keep in mind, but it's not always true. If the claim that you are advancing implies some kind of evidence, then failing to find that kind of evidence is evidence that that thing does not exist. Take, for example, the idea that there might be plesiosaurs in Loch Ness. Well, plesiosaurs had bones. That implies that there should be bones littering the loch. Well, they've dredged the loch to see if there are any monster bones down there, any plesiosaur bones, and there aren't. That goes to the truth of the claim.

**Do you ever encounter people who say, "No, I saw it!"**

DL: Oh yeah. I have a lot of sympathy for that. If you have the experience of seeing something with your own eyes, it's natural that should trump my "talking head" skepticism and Don's arguments about why that's probably not so. But there's only so much I can do with your personal experience that I did not share. I accept that it's compelling to you, but it cannot be as compelling to me.

DP: By and large, all of the evidence for these really strange cryptids is from eyewitness testimony. People are fooled by their senses, especially sight, because we are notoriously bad witnesses. One of the sightings of the Yeti, or the abominable snowman, turns out to be a rock outcrop. The guy saw it move the first time and then he had to leave. He came back finally a year later--after his sighting had been all over the media--and it turns out that it was just a rock he was shooting pictures of.

**What do you think the connection is between people believing in cryptids and the level of scientific literacy among the general public?**

DP: Lately cryptozoology has been connected to creationism in a lot of ways. People who actively search for Loch Ness monsters or Mokele Mbembe do it entirely as creationist ministers. They think that if they found a dinosaur in the Congo it would overturn all of evolution. It wouldn't. It would just be a late-occurring dinosaur, but that's their mistaken notion of evolution.

Science believes that if there is not any history behind bigfoot or cryptids then they can't exist. We find new species all the time and also creatures that were said to be extinct a million years ago. I guess it all come downs to either a carcass or some skeletal remains in order for science to say that they were wrong. Science has been wrong about of lot of things that they said doe's not exist. Like I said I will not be the one trying to prove it to science, just to myself. If they can't except DNA then what is DNA for to prove? It is used almost everyday in convicting criminals so that would mean that the government trust the DNA science. They just do not trust it when it comes to bigfoot? For now science is calling the people who have these sightings and encounters liars. Even respected people even in government have had sightings but they must have had a mistaken identity. We know what we saw, no you know what you think you saw. We don't care if science believes it or not because the only people we need to prove it to is our selves. I do not work for science I do not get paid by science so until then I do not need to prove it to science.

If the government doe's not believe in bigfoot then why are they willing to put us in jail? This came out of California-based paranormal group that want to kill a bigfoot.

If California-based paranormal experts do succeed in killing Bigfoot – the possible Silverton resident – local law enforcement promises to throw the book at them.

Earlier this month, the three hosts of "Paranormal Central" on Dark Matter Radio set parts of the Internet alight when they determined a 2008 photograph – that allegedly captures Bigfoot cavorting in Silverton alongside the Durango & Silverton Narrow Gauge Railroad – might be authentic.

"We don't think it's a kid with a hoodie and a skateboard," said host Jeff Gonzales in a telephone interview Friday.

On the same radio broadcast, Gonzalez and fellow hosts Allen Thomas and Danny Valderrama announced their intention to kill Bigfoot.

Since The Durango Herald published a story Wednesday about the podcast, national media outlets published their own stories. Twitter erupted with innumerable tweets about the veracity of the claims. And numerous others on the Internet, such as the SasquatchChronicles.com, weighed in on why the story and photo resurfaced after the 2008 sighting.

On Friday, Gonzalez said, "We're not ashamed to say it. We came out a few years ago, saying, 'We need to shoot Bigfoot.'" To prove the existence of Bigfoot, scientists will require a body. Then they'll be able to identify Bigfoot DNA and go from there."

The assassination attempt could get financing. The Fox network picked up "Paranormal

Central," which soon will air on TV affiliates in California.

Gonzales said in the last 10 years, Bigfoot has been spotted at least eight times in Silverton. In 2003, La Plata County resident Vi McCoy wrote a letter to the Herald claiming she saw "Sasquatch" while she was riding the train. He is "a solid part of our history with so many people witnessing his presence," she wrote.

"If you are fortunate enough to see this creature, be sure and share it with the newspaper. 'Much thanks to the creature – whomever he may be!'" McCoy wrote.

Local law-enforcement agencies warned that they would come down hard on anyone who bumps Bigfoot.

San Juan County Undersheriff Steve Lowrance said, "As far as our statement on Bigfoot, obviously, we can't discount the existence of such a creature in Silverton."

But, he said, if the people from "Paranormal Central" "do come here and kill Bigfoot, definitely, we'll investigate, working closely with (Colorado Parks and Wildlife).

"We could bring conspiracy charges if there are enough people involved and they took substantial steps in killing such a creature," he said.

Sixth Judicial District Attorney Todd Risberg said while offing Bigfoot would be a criminal act as well as ethically reprehensible, he didn't yet know how he would charge such a slaying.

"If Bigfoot is a human, it would be murder. If Bigfoot is an animal, that might be more of a Division of Wildlife question," he said. "I don't know what Bigfoot is."

Mark Esper, editor of The Silverton Standard, said bringing murder charges might be more appropriate.

"There's a lot of people in Silverton who might be related to Bigfoot. We haven't done the genetic testing yet," he deadpanned.

He said the federal government ought to be doing more to protect Bigfoot from hunters.

"Look – I don't have much of a stake in this. At the Standard, we're very proud of our Bigfoot coverage – but Bigfoot doesn't subscribe or advertise," he said.

"This is prime Bigfoot habitat because of the climate. Given the threats, I don't think the (Environmental Protection Agency) or the Bureau of Land Management is taking Bigfoot protection seriously enough. A lot of people in Silverton want Bigfoot on the endangered species list," he said.

In an email, Joe Lewandowski, spokesman for Colorado Parks and Wildlife, said, "Instead of 'searching' for Bigfoot, these folks should make themselves a nice big bowl of popcorn, slip into their footie jammies and watch the 1987 film 'Harry and the Hendersons.' The movie will give them everything they'll need for an enjoyable evening in the cozy environment of their own homes."

Risberg offered the people at "Paranormal Central" graver advice.

"When I was a judge in Silverton years ago, a waiter used to run by the train in a monkey suit," he said. Anybody planning an attempt on Bigfoot's life "better be careful what they're shooting at."

Photo of the California bigfoot 2008

# Why Government Officials Want to Keep Bigfoot/Sasquatch a Secret

Updated on September 21, 2016

I was browsing around, catching up on my reading, when I came across a couple of questions that were asked by CJ Sledgehammer. Does Bigfoot exist? Why would the government officials want to keep the discovery of Bigfoot a secret? This got me to thinking...

I broke out my bag of sunflower seeds. Snagged a cup of coffee laced with a heavy dose of cream and sugar, and set about working on my theory of "Does Bigfoot exist? Why do I think it is such a secret?". I dusted off my folders, notes, and dug around the net, looking for evidence of the possibility of Bigfoot existing. The Bigfoot evidence was there, but it was only a matter if one wanted to believe it or not.

I thought I would begin with accounts of Bigfoot stories.

# Pitt Lake British Columbia in Canada

**Pitt Lake British Columbia -**

Pitt Lake, British Columbia, Canada

## A man claimed to be kidnapped by Bigfoot.

In 1924, a trapper by the name of Albert Ostman, claimed to have been kidnapped by a large male Bigfoot. He was then taken back to their camp and held for a week before he could escape. He believed he was taken as an unwilling bridegroom for the Bigfoot's daughter.

This is just one of many stories where people believed people were kidnapped to be bred with Bigfoot. It is claimed that there is an area in British Columbia where 21 people have disappeared. (The Pitt Lake area in South Western BC is one such area.) No one knows for sure if Bigfoot is behind it, but many outdoorsmen avoid the area.

## A female Bigfoot was used as a sex slave.

In 1962 a zoologist named Boris Porshnev heard and studied a story of a hominoid named Zana. Zana was a wild woman (Bigfoot) that was captured in Russia. (She died sometime in 1880's or 1890's and was believed to be buried outside the village of Tkhina in Russia.) She was taken to a village where she was held captive, until she became tame enough to do simple chores like a slave. She never learned to talk, only making inarticulate sounds and mutterings. She was described as 6.6 feet tall, having black skin, with her body being covered in reddish-black hair. After being mated with many of the village men, she gave birth to four children—two males, two females-- that lived. The villagers took the infants to raise them when they realized that her previous children died, because she washed them in the ice cold water from the stream. Since her remains couldn't be found, permission was given to exhume the remains of Zana's youngest son, Khwit, which were dug up and studied by Igor Bourtsev. Reports of the study were published in 1987.

More information on this story will be provided in the links below.

## Everything you know is wrong.

## The theory of Lloy Pye

Lloyd Pye has been doing research for years to prove that Bigfoot is part of a hominoid family, and that humans are a genetic mutation. (A result of alien DNA being spliced together with Bigfoot DNA. Is it a possibility? Yes, there are some that think so. Do we know that to be fact? Well, we may soon find out once Melba Ketchum finishes her DNA paper she is currently writing on Bigfoot DNA, gathered from 2 bodies and other specimens.)

In his book called, "Everything You Know is Wrong", Lloyd Pye states that he believes there are four types of Bigfoot. The Bigfoot/Sasquatch, Abominable Snowman/Yeti, Alma/Kaptar, and the Agogwe/Sedapa. He gives examples how Bigfoot could be the Neanderthal man, not just by the shape of the skull, but the position of the limbs and its walk. He claims that they are more suited for this planet than humans. Humans on the other hand are a very fragile creatures. Our bones are thin, eye sockets small-which makes it hard for humans to see in the dark, and the humans walk unbalanced in uneven terrain. According to Lloyd Pye, there is not a single human bone in the so-called pre-human fossil record.

More information on this subject is provided in the links below.

# What if?

What if the Neanderthal man didn't go extinct as predicted 30,000 years ago, but was simply shoved to live in areas that were unappealing to humans such as jungles, swamps, or deep woods? This would explain the stories of the upright, hairy, walking humanoids...aka Bigfoot.

Remember the stories I posted above? One begs to question, how is it possible for a Bigfoot to mate with a human? Take a look at the genetics of a mule. A mule is produced when a female horse mates with a male donkey. All male mules are born sterile, but occasionally a female mule will have estrus cycles and therefore can produce an offspring. But, nothing can be established with certainty, until it is understood how the genes are lined up. The same applies to Bigfoot females mating with male humans or vice versa. More evidence is needed on the hominoid genome to compare them with human genome. Scientists such as Melba Ketchum, Dr. Jeff Meldrum, and the late Dr. Grover Krantz are doing extensive groundbreaking research into this, along with other groups such as the Erickson Project, the Olympian Project, the BFRO, and NABS (North American Bigfoot Search).

# Battle Mountain

**battle mountain nevada -**

Battle Mountain, NV 89820, USA

# Bigfoot deaths

A story that stuck in my mind was about a claim of an elaborate cover up dealing with a Bigfoot that was caught in a fire. It was 1999, at Battle Mountain, Nevada, that one of the largest fires at that time broke out. An anonymous government employee witnessed an injured Bigfoot being captured by the firefighters. Officials were called along with medical personal. The Bigfoot was tranquilized and treated then transported to an unknown location believed to be in Northern Idaho. All parties involved were told not to talk about the incident and the story was covered up. The government employee passed on the information to Bigfoot investigators, while a science teacher and acclaimed author, Thom Powell, came forward to state that he believed the story to be true.

Anyone following the Bigfoot news recently, has been watching to see what will become of the two Bigfoot that were killed by a well known big game hunter. Speculations have been made who has control over the bodies and where they are hid. There was even a claim as to how much the Bigfoot steaks would be worth. In one report, I saw a sale price of $10,000.00, although Melba Ketchum was given one for her DNA analysis. (To say that I was a bit floored would be an understatement.) There have been rumors circulating that the parties involved will be coming out with undisputed physical proof that Bigfoot exists, along with a documentary of a habituation site where clear footage of Bigfoot were taken. I will post more information on this in the following month or so.

# Why?

This brings me back to the haunting question...why the cover up?

To admit that Bigfoot was real would be admitting that Science was wrong. There would be a lot of time and money invested in re-writing history books, documentaries, and other historical documents. Documents-- such as newspaper reports, field reports, police reports-- that have been published, then pulled from the public about Bigfoot, will need to be re-submitted as evidence, as a further admittance of Bigfoot's existence. The Battle Mountain fire would be one of many cases that had a noticeable coverup where evidence was hid from the public.

If the evidence of Bigfoot was miss categorized , there would be a need to go back and restudy that evidence to be properly categorized, reports re-written, and a new placement in the system. That alone would be a financial strain on all parties involved, which may result in a further cover up to avoid the time consuming and expensive task.

To show and admit proof of Bigfoot would involve re-programming the way society thinks and reacts to this species. There are states that are passing laws to protect anyone from harming or shooting a Bigfoot. Many in society do not realize how often and how many species of Bigfoot are killed since this information isn't publicized. I will post a link below on further proof of these killings. One report states that as many as 32 Bigfoot have been killed over the last 125 years. It is believed that there has been a possible government cover up on some of these incidences up since 1968.

What, if any, industries would be effected by the publicity and physical proof of Bigfoot?

The logging industry would feel threatened because they may face the possibilities of being shut down to further protect the Bigfoot environment. This would create a loss for jobs, resources we depend upon, and revenue from those resources.

# 911 call on a man that spotted Bigfoot

It would also have an impact on people that hunt, fish, and camp in Bigfoot's environment. This may become a new challenge for those parties that enjoy big game hunting, and see this as another trophy to mount. Others will feel the need to place Bigfoot on an endangered species list. Parties camping or fishing with young children may feel threatened by the idea of Bigfoot kidnapping them.

Lastly, outdoor sports such as ski, snowmobiling, and snowboarding will be effected. The loss of revenue alone from shutting down such games would leave a huge impact on society's wallet.

In a nutshell, it boils down to this...the cost and impact of Bigfoot being alive and living next to us is enormous. **Who wants to foot that bill?**

The Government confirmed bigfoot.

There has been many questions about bigfoot, is it real? is it all a hoax?
It seems that back in the 70's the u.s. government kind of confirmed that bigfoot was indeed a real creature. While some will always reject any info I could post or just say it's all a hoax, this information should be consider and looked at. This is a big post so please take the time to read it and comment.

In July 1975, The Washington Star-News report:
Though conceding that his existence is "hotly disputed," the Army Corps of Engineers has officially recognized Sasquatch, the elusive and supposed legendary creature of the Pacific Northwest mountains. Also known as Big Foot, Sasquatch is described in the just-published "Washington Environmental Atlas" as standing as tall as 12 feet and weighing as much as half a ton, covered with long hair except for face and hands, and having "a distinctive human-like form." The atlas, which cost $200,000 to put out, offers a map pin pointing all known reports of Sasquatch sightings, and notes that a sample of reputed Sasquatch hair was analyzed by the FBI and found to belong to no known animal.

The last piece of information, that concerning analysis of hair led to the investigation by Peter Byrne of the Bigfoot Research Project in Mount Hood, Oregon. Byrne wrote to the FBI stating: "Will you kindly, to set the record straight, once and for all, inform us if the FBI has examined hair which might be that of a Bigfoot; when this took place, if it did take place; what the results of the analysis were."
The FBI replied to Byrne's request as follows: "Since the publication of the "Washington Environmental Atlas" in 1975, which referred to such examinations, we have received several inquiries similar to yours. However, we have been unable to locate any references to such examinations in our files."

The FBI did, however, follow-up with a Dr. Steve Rice, who was editor of the Army Atlas. In an official report, the FBI states: "After checking, Dr. Rice was unable to locate his source of the reported FBI hair examination."

The Army Atlas and other information relative to the FBI was obtained by Jody Cook under the Freedom of Information-Privacy Act. Jody wrote to the FBI and requested all information relative to Bigfoot. He received documentation and it appears there is always a little "mystery" whenever the FBI is involved in anything. Attached to the file sent to Jody Cook was a standard pre-printed "Dear Requester" form. Curiously, a box on this form which states, "See additional information which follows" is checked. At the bottom of the form the following information was manually typed-in:

"Enclosed are previously processed documents which relate to "Big Foot." The enclosed are the best copies available. Serial 4 is missing from file 95-213013, the file where your release originates. Our effort to locate that document was not successful. It is possible that the number 4 was missed during the original serialization of the file."

So what was in Serial 4? Will we ever know for sure?
Certainly the FBI would be a little more efficient in their filing procedures to omit this section. We have, however, learned that analysis of hair samples as indicated in The Washington Star-News article definitely took place.
George Clappison did extensive research on this incident and was referred by the FBI to the ex-head of their Hair and Fiber Unit. This person, who now runs his own private laboratory out of his home, was in charge at the time the hair samples were submitted to the FBI.

He told Clappison that the analysis was done after hours on employees' own time. He further stated that no written reports were prepared on the analysis. In discussing the whole situation with the current head of the FBI Hair and Fiber Unit, Clappison asked if the unit would now consider analyzing other hair samples. The current manager agreed to perform an analysis, however, he informed the unit would not respond in writing on their findings.

Now back to the Washington Atlas :
In the Washington Atlas, it describes Bigfoot as being a very large animal of 8 to 10 feet tall, weighing up to 1000 pounds with feet measuring up to 24 inches long. The width of the footprint is up to 10 inches wide.

More Depts confirm –
Here's another release –
Dept. of the Interior, Fish and Wildlife Service news release
December 21, 1977 –
Keith Schreiner is Associate Director of the U.S. Fish and Wildlife Service. The service is the Government agency with responsibil…ity for protecting endangered and threatened species…The key law in preservation of a species is the Endangered Species Act, which pledges the United States to conserve species of plants and animals facing extinction. This broad, complex law protects endangered species from killing, harassment, and other forms of exploitation. The Act prohibits the import and export of, and interstate commerce in, endangered species. American citizens cannot engage in commercial traffic in endangered species between nations, even when the United States is not involved. Scientists wishing to study endangered species are required to have a permit issued by the

U.S. Fish and Wildlife Service.

Schreiner acknowledged, however, that a good deal of international cooperation would be needed if extremely rare species were found abroad. And finding one on U.S. soil would pose serious problems too.

The key law in preservation of a species is the Endangered Species Act, which pledges the United States to conserve species of plants and animals facing extinction. This broad, complex law protects endangered species from killing, harassment, and other forms of exploitation. The Act prohibits the import and export of, and interstate commerce in, endangered species. American citizens cannot engage in commercial traffic in endangered species between nations, even when the United States is not involved. Scientists wishing to study endangered species are required to have a permit issued by the U.S. Fish and Wildlife Service.

But before a creature can receive protection under the Endangered Species Act, a number of actions normally must occur which involve recommendations from the public, scientists, and State and foreign governments where the species exists.

The first of these would he the species' formal description and naming in a recognized scientific publication. In addition, if it were a U.S. species, the Governor of the State where it was found would be contacted, as would the officials of foreign governments if it were found outside the United States. Only after much information was collected could the Service make a formal determination as to whether the species should be afforded endangered or threatened status.

For the Loch Ness monster, the first step has already been taken. Last year, a highly respected British journal published a description and proposed the name Nessiteras rhombopteryx, meaning "awesome monster of Loch Ness with a diamond-shaped fin."

Bigfoot, also known as Sasquatch, is purported to be an 8-foot, 900-pound humanoid that roams the forest and wilderness areas of the Pacific Northwest. One "eyewitness" described an obviously female Sasquatch as a "tall, long-legged, gorilla-like animal covered with dark hair and endowed with a pendulous pair of breasts." It, too, has been described in publications and given a scientific name. In fact, so many people were stalking Bigfoot with high-powered rifles and cameras that Skamania County, Washington, is prepared to impose a fine of $10,000 and a 5-year jail term on anyone who kills a Bigfoot. The U.S. Army Corps of Engineers even lists Bigfoot as one of the native species in its Environmental Atlas for Washington. This year the Florida and Oregon legislatures also considered bills protecting "Bigfoot" type creatures. A Bureau of Indian Affairs policeman has 18-inch plaster cast footprints of the "McLaughlin monster," a Bigfoot-type creature he saw last month in South Dakota.

Under U.S. Law, the Secretary of the Interior is empowered to list as threatened or endangered a species for 120 days on an emergency basis. For endangered species in the United States, the Secretary can also designate habitat that is critical to their survival. No Federal agency could then authorize, fund, or carry out any activities which would adversely modify that habitat.

So long-term Federal protection of Nessie or Bigfoot would basically be a matter of following the same regulatory mechanisms already used in protecting whooping cranes and tigers.

So Now the question becomes, Are there any record of the military encountering a bigfoot? and the answer is yes. Consider the following

Fort Stewart
Savannah, Georgia
In December of 1995, a Sargent known only as "R.B." was training in a remote area with two other infantry squad leaders and two privates. Their assignment was to reconnoiter another unit located several miles distant. To avoid detection in the growing darkness, the men waded into chest deep swamp water for several hundred yards, guided by a map and compass. After leaving the swamp, Sgt. R. B. noticed that for the next 15 minutes a "shadow" would approach to within twenty-five yards of the team, but just when they expected to catch sight of their unidentified pursuer, it would break off. In an effort to elude whatever was trailing them, Sgt. R.B., another sergeant, and a private halted while the other two men continued. Five minutes later the other soldiers followed, but the mystery creature was not fooled. Sgt. R. B. wrote: "We could hear something or someone following us… (it would) go around us, fade off, and then we would hear it coming again… (it) started weaving in and out of the two groups like a figure-8… it came between the 2 groups, around them, toward us, like it knew exactly where we were even though we couldn't see it."
When the reunited team was 400 meters from the opposing unit, they established a patrol base. Sgt. R.B. and another sergeant would reconnoiter their objective, then return in two hours, when a second pair would start out. After covering 200 meters the men heard their nemesis again. They crossed an open area, then fell to the ground watching their rear. After five minutes they heard a "scream roar." Sgt. R.B. recorded: "This is when it got really weird… a minute later it screamed again and then we heard what sounded like a huge rotten tree falling, and brush breaking. That was the last we heard and the last I wanted to hear."

A Second Encounter-
The second notable encounter occurred at 2 am. on November 12, 1998, "Barry" was gunner on an armored tracked vehicle that was maneuvering on dirt tank trails in the swampy northeast portion of Fort Stewart. Barry was scanning for simulated targets through a thermal sight while the driver was driving using a night vision device. At one intersection Barry scanned the track and "observed something come out of the vegetation" 50 yards away, entering the brush on the opposite side. It crossed the 15 to 20-foot wide lane in three easy strides. The creature ignored the approaching tank, keeping its head and torso looking straight ahead.
Barry, who had trained with night vision equipment, estimated the creature's height to be 8-feet tall. He also concluded that this hairy entity was unclothed due to the fact that it: "Appeared through the thermal sight to be one constant color from head to foot."
Barry had watched documentaries and read books about Bigfoot as a kid and he immediately thought this was one. Through the Bradley's internal communications system he reported the sighting, but the spooked commander ordered the driver to "punch it" and the vehicle quickly exited the area.
Barry's story swiftly spread through the camp, with the soldier taking "some heavy ribbing over seeing Bigfoot." Later another soldier approached and confided that his father — while stationed at Fort Stewart in the 1960s — had seen an identical

UNCLASSIFIED animal in that area.

Other encounters –
There was a more recent report at Ft.Lewis Army base in aug,19,2008. The report was from
Jeffery Fullerton A SFC MIL USA TRADOC USAREC.

FT.Lewis,Washington May 1984

"I was a sergeant of Military Police at the time of the incident, stationed at Ft. Lewis, Washington.

I went to investigate a report of a disturbance within the tree line, near the post stockade. This was early May 1984, approximately 0300-0400 hrs (the bars had closed shortly before the call).

The MPDO heard strange cries from within the forest and wanted it checked out. I went in one direction while the K-9 unit went in another to sweep the area.
We planned to meet up again at an old railroad spur, not too far into the trees. I saw nothing nor heard a sound until the K-9 apparently made contact with something, then I heard five distinct pistol shots, at which point I heard a deep, guttural, growl building into an extremely high pitched howling (I'd never heard anything like that) and the sounds of something large crashing through the thick brush and foliage in the area. (Important to note that I was too young and gung-ho, i.e. stupid, to be scared)!
I was armed with my issue .45 and 12 ga. riot gun and continued to the rendezvous point, hearing nothing further.
At the spur (sort of sunken with high berms), I went up the far side and halted at the edge of a large meadow. The captain was already there at the rendezvous, having not seen anything either. Neither of us knew that the K-9 unit had fled the forest on the trail of the dog. We were about to head back when I caught movement in the adjacent tree line off to my left. I could plainly see a large dark shape
mocking along the southeast edge of the meadow, but still within the tree line. (Mt. Rainier was Southeast of my location and this was the direction of movement.) It appeared to be a bear.

I held my weapon at the ready. When the object turned and came out and into the meadow, it was approximately 35 yards from our position moving from left to right. It did not register at first, as I nearly pulled the trigger, but something didn't look right. Bears walk on all fours; this bear was clearly walking on two legs.
It was getting on towards morning as false dawn was in evidence. I could see well enough, but not as to clearly make out facial features.

All I can figure is the wind must have shifted for the creature stopped and turned its head and looked directly at me. It turned its whole body and just stood there looking at us, arms by its sides. The creature was not threatening us at all so I lowered my weapon and did not open fire. I remember the head moving slightly from side to side… it did not move closer and neither did we.

We stared at each other for less than 2-3 minutes (maybe less) ultimately it resumed its original direction and walked away looking back once but kept going, disappearing into the opposite tree line.

The animal was covered with short, dark hair, massive arms and shoulders, probably 7 1/2 to 8 feet tall, the neck was not evident, the head bullet-shaped, the face appeared ape-like (though due to poor light conditions I could not see facial features clearly) but I can say that it definitely was not Elvis nor the Pillsbury dough boy. I estimate the weight at close to 500 lbs., (as tracks we found later would attest). I estimated the size by comparison.

My son likes WWF Wrestling and I met Andre the giant and shook his hand. If he is really 7'4", then this creature was taller yet as I am 6'5" and recall the size difference between Andre the giant wrestler and me. At the time, it was said we were either drunk, crazy or on drugs as I recall the harassment I had to endure.

I can't imagine a person so bored with life or in need of attention to purposely set himself to look like a fool. I'll tell you now; no official report was ever made. I valued my army career more than a few moments of limelight.
As I said, I saw an ape-like creature. I never said I saw bigfoot nor had a doughnut with a sasquatch. But I'm 43 now, retired military. I cant prove any of this, all I can tell you is I know what I saw, it no longer matters who believes me or if nobody does.

Back in 1993, I went to Caro, Michigan to the Michigan-Canadian Bigfoot Center, Wayne King's home. I told him of my incident
and he played an audio cassette of a supposed bigfoot for me.
The scream was exactly as that one I heard so long ago. I must say further, that the creature I saw did not appear to act as though it had recently been shot at.
Reported by Mike Coppola


Oregon Encounter-
Todd Neiss of Oregon was witness to three Bigfoot in 1993 in Oregon while on military duty.
Todd was a sergeant in Charlie Company (1249th Combat Engineers) and while training on a mission of Combat Engineering witnessed 3 sasquatch, he said this "In the middle stood, what I assumed to be, the alpha male of the group; as it towered a full head above the two creatures that flanked it. I would estimate it to have stood approximately nine feet high, with the flanking creatures approaching seven feet in height"

There are many more stories about the military and sasquatch running into each other but I'll stop with the stories for now.

Lets overview a little:
Talked about bigfoot – Army Corps of Engineers , Dept. of the Interior, Fish and Wildlife Service, FBI , The Washington Star, Washington Atlas .
Military Reports –
Many reports and several coming from the Washington area. Others from Oregon and Georgia just to name a few.

So in closing it appears that many government agencies, including the military, have already accepted that bigfoot is a real creature.

## Engineers Corps Book Recognizes 'Big Foot'

SPOKANE, Wash. (AP) — Sasquatch, the elusive legendary creature of the Pacific Northwest mountains, has been officially recognized by the Army Corps of Engineers.

Though branded as a myth by some, Sasquatch is described in detail in the "Washington Environmental Atlas," a $200,000 Corps project designed to assist government and private planners.

The book says Sasquatch, also known as Big Foot, stands up to 12 feet tall, weighs up to 1,000 pounds and strides up to 6 feet.

The animal is "reported to feed on vegetation and some meat and is covered with long hair, except for the face and hands, and has a distinctly humanlike form," the atlas adds.

It says the beast is agile and strong, but so shy that it leaves "minimal evidence of its presence."

The corps acknowledges that the existence of Sasquatch is "hotly disputed." It adds some persons believe that "not one piece of evidence will withstand serious scientific scrutiny."

However, the atlas also provides a map purporting all reports of Big Foot sightings. And it notes that hair claimed to be from Sasquatch was found on FBI analysis not to have come from man or any known animal.

"If Sasquatch is purely legendary, the legend is likely to be a long time in dying," the atlas says.

The atlas, prepared over a three-year period, includes sections on Washington plants, archeological sites, rivers and lakes of environmental interest, geologic features and historical and contemporary points of interest.

There you have it conflicting beliefs of the legend, if it doe's not exist then why do so many government agencys keep bringing there claims to the public and are willing to put us in jail if we shoot one? Putting up bigfoot signs in our national parks and Forrest. Some states have even made it illegal to shoot one. So then I guess science has some catching up to do.

www.ingramcontent.com/pod-product-compliance
Lightning Source LLC
Chambersburg PA
CBHW041944240526
45473CB00033B/516